The summer day is hot
and still. The sky gets dark
quickly. The world waits for
the storm to start.

Soon a cool rain wets the earth. The dry corn drinks it up. Happy ducks quack in the rain.

Kids at play quit their
sports. They play word games
on the porch. Or they play
inside. But other people work
in the rain.

It rains quite hard for a long time. The raindrops form little puddles. The little puddles get big.

There is a flash of light.
Thunder cracks and booms!
Rain fills the ponds and lakes.

Cars honk their horns.
Ships wait in port. Will this
storm be short or long?

At last the rain stops.
The sun shines in the sky.
The world seems bright
and fresh.

Kids play in the cool air.

Soon the puddles will dry up.

The storm has ended.

The End